28

rritory

104
105

98

90

65

92
Cairns

40

94

Great Barrier Reef

Queensland

108
109
Rockhampton

19

72

52

16

50

BRISBANE

48 Surfer's Paradise

outh Australia

63

66

76

86

88

25

85

17

29

81
A

81
B

New South Wales

45

43

82

83
Newcastle

ADELAIDE
95

68

57

57 Victoria
106

91

87

57 SYDNEY

CANBERRA
61

84

58

60

Mount Gambier

MELBOURNE
56 70

71

111

Wilson's Promontory

Tasmania

62
HOBART

Key to Location of Plates

The numbers appearing on this map relate to the location where the subject of the picture was photographed. Most animals and wild-flowers have a wide range of distribution and some occur throughout the continent. Distribution details are given in captions to the colour plates for most species.

The Nature of
Australia

The Nature of Australia

Michael Morcombe

REED

First published 1980

This edition published 1985 by
REED BOOKS PTY LTD
2 Aquatic Drive Frenchs Forest NSW 2086

National Library of Australia
Cataloguing-in-Publication Data

Morcombe, Michael.
  Nature of Australia.
  Previously published: Adelaide: Rigby, 1980.
  Includes index.
  ISBN 0 7301 0050 2.
  1. Natural history – Australia – Pictorial works.
  I. Title.
574.994

Printed in Singapore by Tien Mah Litho Printing.

Contents

Giant Karri Trees, south-western Australia

In the rain-drenched south-western corner of the continent the white and salmon-splashed columns of the Karris soar smooth and straight towards the sky, rising 200 feet or more to the first limbs, then higher again until lost in their own canopy of leaves almost 300 feet above the ground. Below, in a cool shaded world where only an occasional shaft of sunlight moves across the forest floor, are the Karri wildflowers, the flowering shrubs and creepers of this unique type of forest environment. The Karri, *Eucalyptus diversicolor*, is one of the world's tallest trees; some areas of prime Karri forest have been preserved in national parks.

Soaring white trunks and branches of huge Karri trees against deep blue sky; salmon pink and cinnamon-toned butts of Brown Mallet trees shedding their bark; rough-textured mossy trunk and green, gold or red translucent leaves of an Antarctic Beech—these are sights as fascinating as the most brightly colourful flower.

The term 'flora' encompasses all vegetation native to Australia, not only wildflowers (a word in common usage which generally refers to plants with conspicuous flowers) but also trees, palms, ferns and grasses. Each and every one of these is not only an intrinsic part of the nature of Australia, but is itself of great beauty.

The age-long isolation of this continent has given sufficient time for evolution to proceed along many paths, with each plant becoming more and more changed compared with others which once were similar but which came to inhabit different environments. As a result, very few species have remained identical to plants of any other continent. An exceptional proportion of our plants are endemic, that is, they are found only in Australia.

For this reason conservation of the flora is crucial; once a species becomes extinct in Australia it is simultaneously lost to the entire world.

Unlike the flora of the northern hemisphere, our evergreen forests have been able to occupy almost every part of the continent. We have evergreens (mostly *Eucalyptus* species) on mountain slopes that are covered with snow much of the year. Evergreens are scattered almost throughout the arid regions and yet they dominate the wet tropical and sub-tropical forests. By contrast, in the northern hemisphere the colder parts have conifers, the milder regions have deciduous summer-green forests, and the arid zones are characterized by succulents.

This invasion of such an extreme range of habitats by the Australian evergreen forests has been made possible only by specialized adaptations giving the trees of different regions a high tolerance to cold, drought or flooding.

The various natural environments of Australia each have their own characteristic vegetation. For example, the dry areas have extensive tracts of mallee eucalypts. Each mallee has many stems arising from a single massive underground rootstock. Around the base of the multiple stems accumulate all the sticks and dead leaves that would otherwise blow away, so that in a thicket of mallee much of the ground has a layer of debris that keeps the soil cool and moist for a long time after rain. Following bush fires, which occur quite frequently, new shoots appear at ground level, and within several years a low mallee thicket has reappeared.

Across huge tracts of country that are even more arid, the *Acacia* scrub, known as mulga, is dominant. There are many species of this scrub, including Witchetty Bush, Myall, Mulga,

Prickly Wattle and Gidya.

Extensive displays of colourful flowers can be seen only after good rains in arid and semi-arid parts, when for several months parakeelyas, everlastings, eremophilas, pityrodias and a great many others will bloom.

In the tropical north and north-east, extending to the sub-tropical east in fertile and well-watered localities, are areas of rainforest, some quite extensive, but many reduced by clearing except in the most mountainous parts.

Unlike the eucalypt forests the rainforests include a great many species, sometimes as many as a hundred different trees per acre. It is in the rainforest that the greatest variety of plant life occurs, with a profusion of epiphytes forming aerial gardens on the limbs of huge trees, as well as palms, ferns and lianas, and the muted greens and browns of lichens and mosses on almost every surface. Most trees have tall, narrow, wide-reaching buttresses, radiating from the lower part of their trunks, a feature which is almost a characteristic of the rainforests. It is thought that these plank-buttresses, like the breathing roots of mangroves, assist the aeration of the tree roots in warm wet climates particularly where the drainage is poor and the soil always wet. The leaves, of a bewildering variety of shapes, have narrow-pointed drip-tips to drain off excess water during the days or weeks of almost incessant rain—a contrast to the water-conserving leaves of the desert trees.

Conspicuous in the rainforests are the huge strangler figs which encase a tree in a network of roots that interlock, thicken, and slowly choke it to death.

These figs begin life as epiphytes, that is, they germinate from sticky seeds deposited by birds, possums or fruit bats on branches a hundred feet or more above the ground. There they take hold, and soon drop their strangling roots, very thin at first, to the ground.

The tropical Australian vegetation is not only rich, but is also varied in character. As well as rainforests there are mangroves, swamp forests, tropical grasslands and savannah-woodlands, hardwood forests, and on some north-eastern peaks, cloud forests.

In south-eastern Australia the flora of the coast and its immediate hinterland is of great interest. In the east this includes the sandstone country of the Sydney basin, as well as the coastal sands of northern New South Wales, south-eastern Queensland, southern New South Wales and south-eastern Victoria. These low-nutrient sands, together with some granite-derived soils in the tablelands, produce an excellent variety of flowering plants. Eucalypts are the dominant large trees, with many small shrubs in the understorey, including such spectacular species as the Waratah and the large-flowered Christmas Bell.

In south-eastern Australia there are the tall wet-sclerophyll

River Red Gum, *Eucalyptus camaldulensis*
River Red Gums, their trunks and branches white against red cliffs and deep blue skies, are characteristic of gorges, rivers and watercourses throughout arid and semi-arid parts of inland Australia. The River Red Gum can be a tree of great beauty in these iron-rust-red gorges of the Hamersley Ranges; it is seen along river beds in Central Australia, and in eastern Australia it grows beside the billabongs and meandering channels of the Murray–Darling basin. It is generally confined to watercourses which flow only intermittently or seasonally. Its foliage is of a refreshing bright green, in a dense shady crown, and its broken limbs soon become deep hollows that are used as nests by parrots and cockatoos. The River Red Gum is one of the most valuable trees of the interior, welcomed by man and bird alike for its shade and shelter.

forests of the mountain ranges, where trees reach gigantic size. The Alpine Ash reaches heights of 200 feet, often forming pure stands at altitudes between 3,000 and 4,000 feet, while the world's tallest flowering plant, the Mountain Ash, grows to heights of 300 feet in the cool moist highland valleys. On a few mountain summits near the Queensland–New South Wales border grows the Antarctic Beech, one of the trees that indicate a past botanical link with South America via Antarctica.

In damp places all through these mountain forests of eastern Australia and Tasmania are the fern gullies, where tree-ferns grow to heights of 40 feet.

But it is in the isolated south-western corner of Australia that the flora reaches its climax in number of species, novelty of form and brilliance of colour. Separated from the plants of eastern and northern Australia by extensive deserts, those of the west have exploded into a bewildering variety of species not found in any other region.

Large groups of such unusual shrubs as those of the genus *Dryandra* (of which the Showy Dryandra is a spectacular example) are confined to the south-west.

Banksias have proliferated, with a great number of large-flowered trees and shrubs typified by the Scarlet Banksia and the Swamp Banksia, while among smaller plants occur such strange forms as the Kangaroo-paws, including one called the 'Cats-paw'.

The wildflowers in this book have been photographed in their natural environments; often their surroundings have been included in the photograph. But a scenic background is not in every case possible or desirable. Sometimes the surroundings had insufficient character, while on many occasions the flower could be more effectively shown by deliberate exclusion of all background detail, thereby concentrating attention upon the inherent beauty of the subject, particularly in extreme close-up studies.

Ancient Wandoo gum, *Eucalyptus wandoo*
'Wandoo' was the Aboriginal name for this widespread Western Australian tree which attains heights of 80 to 100 feet. It is a valuable forestry tree, with an extremely hard and durable timber. This gnarled and hollow old specimen stands in the Dryandra State Forest, an extensive tract of natural woodland and mallee country which is also a refuge of the Marsupial Ant-eater or Numbat. One of the most beautiful of all marsupials, but rare in most other places, the Numbats live in hollow limbs dropped by trees such as this. Although no use for commercial forestry purposes this old Wandoo provides hollow limb nest sites for parrots and possums; it is worth preserving as part of a wildlife habitat as well as for the distinctly Australian character of its gnarled old trunk and twisted limbs.

Above:
Flood-damaged River Red Gum, Central
Australia *Eucalyptus camoldnlensis*
Pushed almost horizontal by some huge flood
rushing through the Finke River Gorge near
Palm Valley, Central Australia, this River Red
Gum has sprouted numerous new stems from
the original trunk. Beyond the river channel
rise the red cliffs of the Finke's long gorge,
south-west of Alice Springs.

Opposite:
Baobab or Bottle Tree, tropical Australia,
Adansonia species
Baobabs, characteristic trees of tropical northern
Australia, are famed for their huge swollen
bottle-like trunks, which vary greatly in shape.
Although the name 'bottle-tree' comes from
their bottle-like trunk profile rather than from
any water-holding capacity, it is sometimes
possible to obtain small quantities of water from
hollows at the bases of the branches. Baobabs
are deciduous, the leaves turning yellow and
falling off in the dry season, exposing large
globular fruits from which the Aborigines
obtained a floury paste. Baobab nuts, dried
and carved in traditional patterns, can be
bought from some of the native missions in the
Kimberleys.

Above:

Brown Mallet, *Eucalyptus astringens*

Australia has more than 600 species of eucalypts, ranging from forest giants to stunted mallee types. One of the most variable features is their bark, which may be smooth or extremely rough, exhibiting a great diversity of colour. With some, like this Brown Mallet of the inland woodlands of south-western Australia, a colour change may occur when the bark is shed each year. The Mallet's bark is smooth and silvery-grey, flaking away in sheets and ribbons to expose new bark of a rich salmon colour with patches of rough, dark-brown resinous gum adhering here and there. Where stands of these trees occur, the numerous glossy brown and salmon-tinted trunks and branches rising against a blue sky, and lit by full sunlight, have a unique beauty, purely Australian in character.

Opposite:

Bushy Yate, *Eucalyptus lehmannii*

The flowers of the Bushy Yate are fused together in spherical clusters, from which the long finger-like bud caps radiate outwards. Many flowers, usually green or yellowish, together form very large spherical clusters. The fruits are curious rounded spiked structures most unlike the ordinary gumnut. This tree is small and slender, and very suited to growing in gardens. It comes from the south coast of Western Australia.

Above:

Screw-pine, *Pandanus*

Pandanus, a palm-like tree with stilt roots, characterizes Australia's eastern and northern coastline and the islands of the Pacific. There are several species, occurring on the reef islets, beaches, along tropical freshwater streams and in tropical woodlands and grasslands. These pandanus trees silhouetted against a Pacific sunrise are at Noosa, southern Queensland.

Opposite:

Antarctic Beech, *Nothofagus moorei*

Gnarled and mossy giants of tremendous age, many carrying on their limbs great clumps of staghorn, elkhorn and other ferns, and aerial orchids, the ancient-looking Antarctic Beech trees in Australia are confined to a few mountain tops. They occur on the McPherson Ranges of south-east Queensland, on peaks in New South Wales, and in parts of Victoria and Tasmania. Where the sun breaks through the treetop ceiling of foliage it lights the leaves nearest the camera a brilliant luminous green, or a dying leaf here and there a fiery red. Trees of this ancient genus occur also in South America and New Zealand, and their fossilized pollen has been identified in Antarctica.

Above:

A small wattle, *Acacia pyrifolia*

Growing in the river gorges of the Hamersley Ranges of arid north-western Australia, this wattle is but one of approximately 600 species of *Acacia* native to Australia. A curious feature of the wattles is that on many species the leaves are replaced by phyllodes, modified leaf-stalks which in turn have in many cases evolved a very leaf-like shape.

Opposite:

Cabbage-tree Palms, *Livistonia*

Tall Cabbage-tree Palms are back-lit by afternoon sunlight which casts smoky-blue shadows over the cliffs of Queensland's fascinating Carnarvon Gorge. Although this is a relatively dry region 250 miles inland, coastal flora including the palms, tree-ferns, maiden-hair ferns and elkhorn ferns flourishes in the shaded dampness of the 600-foot sandstone gorges. There are six or seven species of *Livistonia*, known as Cabbage-tree Palms, native to Australia, found mainly near the northern and eastern coasts. Their very large leaves or fronds are fan-shaped, glossy and bright green, making these trees quite ornamental.

Opposite:
Wattle Tree, Northern Territory, *Acacia* species
Growing among scattered eucalypts on the broad grassland plains of the 'top end' of the Territory, this wattle tree has exceptionally large, long cylindrical golden flower spikes borne in great profusion during the dry season: June, July and August. Wattles rival the eucalypts in their almost universal distribution, occurring in most parts of Australia, from mountain forests to semi-desert plains.

Over:
A Rose Cone Bush, *Isopogon latifolius*

Above:
Blackboys, *Xanthorrhoea preissii*
One of Australia's botanical curiosities, the grass-trees are one of the most unusual members of the lily family. Their stems are so often blackened by bushfires that they are known in places as 'blackboys'; the resemblance to a native warrior, when seen silhouetted against the skyline at sunset, is heightened by their tall spear-like flower spikes. There are some fifteen different species of these ancient plants, occurring in all states but not in the Northern Territory. The hardy, fire-resistant blackboys grow to a height about twenty feet, but growth is extremely slow; other species remain dwarfed, never developing a stem to lift the leaves much above the ground.

Above:

Cats-paw, *Anigozanthos humilis*

Although at first glance seeming to be a stunted and unusually-coloured specimen of the well-known kangaroo-paw, the Cats-paw is a distinct species, growing usually no more than eighteen inches tall and often standing as little as six inches above the ground. It is very widespread in south-western Australia, where its colour varies from yellow through bright gold and fiery orange to rich brownish reds. The flowers bear a superficial resemblance to the paw of a cat with claws extended.

Opposite:

Large-flowered Christmas Bell, *Blandfordia grandiflora*

From among tangled undergrowth beside the Dandahra Creek in the Gibraltar Range National Park, north-eastern New South Wales, rise the tall stems of a particularly large-flowered form of the well-known Christmas bells. Its broadly dilated floral tubes are almost three inches long, giving it a more bell-like appearance than the other species of *Blandfordia*. The Large-flowered Christmas Bell occurs on the coast and tablelands of eastern New South Wales from the Hawkesbury River northwards to Queensland, flowering from November to January.

Above:

Golden Featherflowers, *Verticordia nitens*

On Western Australia's coastal sandplains, between Perth and Geraldton, the display of wildflowers each year reaches a spectacular climax in November and December with the blooming of the Golden Featherflowers, or Morrison, as this species is also called. This is banksia woodland country, where the slender *Verticordia nitens* grows in huge numbers, each plant consisting of a straight thin unbranching stem three to six feet high, supporting a huge crown a foot or more across, of tiny feathery-edged flowers. In this same region a number of other spectacular wildflowers come into flower at this time of year—the orange-flowered Christmas Tree, the Golden Kangaroo-paw, and the magenta-and-yellow *Pileanthus* shrubs.

Opposite:

Round-leaved Parakeelya, *Calandrinia remota*

Spreading like a carpet across the red sands of Central Australia, these low fleshy-leaved annuals will, by the water-storing ability of their leaves, continue to flower for months after the last of the winter rains. Not every year does the dry heart of Australia put on such wildflower displays as this. More often than not the arid nature of the centre is all too evident, with few if any wildflowers, as year after year the winter rains fail to reach this far inland. Drought and near-drought may prevail for three, four or five years at a time. But when that grim pattern is broken and the first flooding rains sweep in, then the wildflower displays of the Australian deserts are magnificent. This parakeelya, which has flowers about one inch in diameter, occurs in arid parts of the Northern Territory and South Australia.

Above:

Thick-leaved Mallee, *Eucalyptus pachyphylla*
Decoratively ribbed red bud caps, which fall
away as the bright yellow stamens of the large
flowers unfold, are the most attractive feature
of this shrubby little tree. The leaves are large,
broad, very thick and of a leathery texture. The
Thick-leaved Mallee is native to the far north
of the Northern Territory and north-western
Queensland.

Opposite:

Waratah, *Telopea speciossima*
Shown in its natural habitat of eucalypt forest
on the Great Dividing Range of northern New
South Wales, this Waratah shrub carries four
perfect flower heads, each of which is made up
of many tiny, grevillea-like flowers clustered
above an involucre of crimson bracts. The
Waratah flower has evolved for pollination by
small birds. The rows of little flowers at the top
of the head open last, so that there is here a very
convenient perch for honeyeaters. The tubes of
the little flowers curve upwards to fit the curve
of the beak of the honeyeater probing downwards into the ring of encircling flowers for
nectar. The crimson colour is conspicuous in
the bushland and is a hue known to attract
birds, while the stems are strong enough to
support the weight of quite a large bird.

Above:

Swamp Banksia, *Banksia verticillata*

Growing by the edge of a small lake in south-western Australia, the Swamp Banksia is shown in typical natural habitat, where the ground is waterlogged, and the tree often surrounded by shallow water in winter. The cylindrical flower spikes, seen from December to May, may be almost a foot in length, and the slender leaves have a conspicuous white undersurface. This banksia eventually grows into a large tree of thirty to fifty feet in height.

Opposite:

Scarlet Banksia, *Banksia coccinea*

Superficially resembling a Waratah, and therefore occasionally called 'Waratah Banksia', the Scarlet Banksia is an open shrub or slender small tree that grows near the south coast of Western Australia. This specimen was flowering in a valley between peaks of the Stirling Ranges; in the background can be seen the jagged outline of Mt Toolbrunup, which rises to a height of 3,450 feet and dominates forested valleys and sandy flats where the Scarlet Banksia grows in profusion. While the main flowering period is October to December, a few blooms may be found at almost any time of the year.

Above:

Gravel Bottlebrush, *Beaufortia sparsa*

The Beaufortia is a member of a genus of sixteen species all confined to Western Australia. The flowers of some beaufortias have a bottle-brush-like appearance, but the distinguishing feature of the group is the branching of the colourful filaments towards their outer ends. The Gravel Bottlebrush is a tall slender shrub sometimes exceeding six feet in height, growing naturally on sandy and gravelly soils along the southern coast, and flowering from January to June.

Opposite:

Everlastings, *Cephalipterum drummondii*

Transforming the barren mulga flats of the interior in August and September, bright yellow papery everlastings extend in every direction through the sparse scrub as far as the eye can see. *Cephalipterum drummondii*, one of a large number of papery-flowered annuals of the family Compositae, has large globular heads often two inches in diameter. Each of these is a compound head consisting of a large number of separate small flower heads crowded together on a single receptacle, each of these small heads being surrounded by long papery bracts of yellow or white. This species occurs in parts of South Australia and Western Australia.

Above:

Round-fruited Banksia, *Banksia sphaerocarpa*
With colours ranging from gold and yellow to
bronze, rust or brown, often with a touch of
violet, this can be an extremely attractive small
shrub. Widespread and common, with flowers
sometimes almost on the ground, it grows on
sandplains and open heathlands, as well as on
clay and gravelly soils of forests and ranges.
Because it has such an extensive distribution
from mid-western to mid-southern coasts of
Western Australia and occurs in such a variety
of habitats, the Round-fruited Banksia may be
found in flower somewhere in the state at almost
any month of the year.

Opposite:

Dampiera lavandulaceae
The *Dampiera* genus has fifty species in Western
Australia, with flowers somewhat resembling
those of the well-known Leschenaultia, and
usually of a distinctive dusky ultramarine colour.
This extensive spread of *Dampiera* is in the open
woodland country of inland south-western
Australia, where the species is widely distributed.
This plant is low, tufted, rush-like, spreading
rather than growing upwards for any height. It
flowers between July and October. The name
Dampiera commemorates the explorer William
Dampier who collected the first species.

Opposite:

Showy Dryandra, *Dryandra formosa*

At an altitude of almost 4,000 feet on a peak of the Stirling Ranges the Showy Dryandra grows wherever it can find a foothold on the windswept ledges and rocky crevices. The large flowers have a metallic sheen, as if constructed of polished gold wire. Foliage is soft, with finely serrated leaves. Flowering time is September to November. Behind the flowers can be seen some of the huge cliffs that form the northern and eastern walls of Bluff Knoll; in the distance are the ridges leading up to other peaks, with Ellen Peak, almost encircled by cliffs, rising above the skyline.

Above:

A small Rose Cone Bush, *Isopogon baxteri*

Another beautiful small shrub of the rich flora of Western Australia's Stirling Ranges, this Isopogon has flowers of delicate colour and exquisite detail; the foliage also is attractive, with decoratively shaped leaves. The tubular pink flowers are covered with fine white hairs that give a frosted effect, while the protruding styles of the opened flowers are deep orange in colour. This autumn and winter-flowering species attains a height of about three feet, and may be found on the rocky slopes of the ranges.

Above:

Firebush, *Keraudrenia integrifolia*

The flowers of this shrub are said to look like small velvety Chinese lanterns, carried in such profusion along the stems that the shrub is a mass of violet-purple, and the foliage is hardly seen. This *Keraudrenia* grows to a height of about four feet, and occurs in the north-west and inland parts of Western Australia. Like the flowers, the leaves have a velvety surface; they are white underneath. The common name, Firebush, derives from its rapid re-growth on land bared by a bushfire.

left:

Hairy Gland-flower, *Adenanthos barbigera*

The beauty of these small flowers is greatest when they are seen against the sun, when they become like glowing coals in the forest under-growth. At close hand their fine white hairs catch the sun, accentuating the illusion of light from within the flower. This is a slender little forest undershrub which grows on the gravelly soils along the Darling Ranges of Australia's western coast, flowering from September to January.

Above:

Scarlet Featherflower, *Verticordia etheliana*
A very spectacular shrub, up to six feet tall,
with brilliantly coloured flowers up to an inch
in diameter, and very small leaves that are
crowded closely along the stems and not in any
way obscuring the flowers. This featherflower
is native to the red sand plains of the inland
west-coastal areas of Western Australia, blooming from September to November.

right:

Woolly Foxglove, *Pityrodia axillaris*
The tubular flowers of this woolly-foliaged shrub
vary from pink to magenta, and are arranged
in very conspicuous fashion up the stems. The
leaves and flower calyces are so densely covered
in fine hairs that they appear silvery, and
provide an effective contrast to the colour of
the flowers. This *Pityrodia* grows on sandy and
gravelly soils of the southern half of Western
Australia, principally across inland areas.

Rainbow Lorikeets, *Trichoglossus moluccanus*
On the dead limb in which their nest hollow is situated a pair of Rainbow Lorikeets preen their colourful plumage before flying off in search of flowering trees and the nectar upon which they feed. One bird has lifted a wing, showing the intense crimson which otherwise can be glimpsed only for an instant when flying lorikeets dash overhead. The Rainbow Lorikeet, one of the largest members of this nectar-eating group of Australian parrots, inhabits the coastal eucalypt forests of eastern Australia from north Queensland to South Australia.

High, thin-sounding notes and bursts of rapid reeling song mark the passage of a family of wrens through the dense scrub. Suddenly they appear, hopping from beneath the skirted grass-trees, their tails held jauntily erect; they pause a moment, giving a tantalizing glimpse of their dainty shapes, and the bright plumage of the males. After stabbing at small insects here and there in the leaf-litter of the forest floor and calling constantly to one another, the wrens fly off, with long tails streaming behind, into the undergrowth again.

Soon all are lost to sight in the forest thicket, but bursts of song and squeaky calls trace their path beneath wattle and bracken.

This typical encounter with some colourful little birds of the Australian bush occurred in the far south-west of Western Australia, in the rain-soaked Karri forest country. In that region those wrens could have been either of two species, Red-winged Wrens, whose song begins with several high-pitched squeaks, or Splendid Wrens, which launch straight into the powerful, reeling trills of their songs. But the males of this party had chestnut shoulder patches, showing that they were Red-winged Wrens.

Birds, secure in their ability to escape by flying, are the most easily observed segment of Australian wildlife. Glimpses of Australian birdlife might include a Yellow-breasted Sunbird hovering like a hummingbird in front of a jungle flower, or a Spinebill Honeyeater probing for nectar in the flowers of a swaying Kangaroo-paw.

A Black-faced Flycatcher skilfully decorating its nest with green mosses until it blends perfectly with its rainforest surroundings; a Rainbow Bird with orange wings flashing against blue summer sky as it darts after a dragonfly; an Azure Kingfisher shattering the reflections mirrored in still water as it plunges into a river pool—these are typical yet fascinating sightings of Australian birds in their natural environments.

Australia has about 650 species of birds; of these, many can be found in no other place. These endemic species are generally those which, by long geographic isolation, are restricted to certain very definite and circumscribed habitats, and consequently are often of restricted or broken distribution.

Australian birds originated in the Old World tropics and oriental region, but since arrival and subsequent isolation here, have given rise to many species that are peculiar to this continent. Some groups are considered to be relic forms—the last survivors of the bird fauna of past ages.

Bird groups that are entirely endemic to Australia include the two species of scrub-birds, the two species of lyrebirds, the whipbirds, the Wedgebill, the bristle-birds, the quail-thrushes, the grass-wrens, the Spinifex-bird, the Rock-warbler, the Pilot-bird and the Mallee Fowl.

The Mallee Fowl is an example of a bird derived from the avifauna of the humid tropics but subsequently adapted to the semi-arid environment. This adaptation was achieved largely by changed patterns of behaviour.

The Mallee Fowl is one of Australia's three 'incubator-birds', which do not hatch their eggs in the usual way by application of body heat, but instead build large mounds (or incubators) of leaf litter mixed with soil. The heat generated by the fermenting of the vegetation is sufficient to hatch the eggs, so long as there is adequate moisture in the mound.

The other two Australian incubator birds, or megapodes, inhabit rainforests where there is abundant moisture. For them, mound-building is an easy way of hatching a very large number of eggs during a long egg-laying season. These birds, the Brush Turkey and the Jungle Fowl, are typical of the numerous megapodes of the tropical islands northwards of Australia. (The Jungle Fowl in fact is found from Indonesia to the Pacific islands).

But the Mallee Fowl is unique in that it is a megapode inhabiting regions which are arid for a large part of the year. The mound-building method of incubation which made life easy for the jungle-dwelling megapodes has become, for the Mallee Fowl, a method involving constant attention and much hard labour.

The turkey-sized Mallee Fowl builds up its huge pile of leaves, sticks and sand in the late autumn. The winter rains begin the fermentation process, and the first eggs are laid in the warm mound early in the spring. As the rains cease the mound becomes dry, and the heat derived from the rotting vegetation diminishes. But with the approach of summer the direct heat of the sun on the mound keeps the eggs warm. However the male Mallee Fowl must attend the mound almost throughout the day, opening it to admit the sun's rays if it is too cool, heaping up the hot sand in the afternoon to retain the warmth well into the night, or covering the eggs with additional layers of sand if they are becoming too hot. Thus he continues through the summer and into early autumn, when the sun's heat becomes inadequate to maintain the warmth of the dry sand. The mound is then abandoned until the first rains initiate the next cycle of activity.

Although belonging to a group evolved in the damp warm tropical jungles, the Mallee Fowl has, through a very complex pattern of behaviour, survived in a far drier climate where temperatures vary from scorching heat to extreme cold.

Many other Australian bird groups have managed to occupy arid as well as humid environments. Among these are the bowerbirds, the kingfishers, the fairy wrens, and the waterfowl.

Bowerbirds are peculiar to Australia and New Guinea; the male of each species builds his distinctive type of playground or

Top:

Crimson Rosella, *Platycercus elegans*
Dashing away from a hollow high on the side of the trunk of a dead tree, a male Crimson Rosella displays its richly coloured plumage. In the background the green paddocks of a New England Tableland farm are darkened by late afternoon shadows. This photograph, taken from a 'hide' on top of a wooden tower, was obtained by use of special electronic flash equipment which gives an intense flash of light lasting only 6,000 of a second, effectively 'freezing' all movement, and capturing details of the actions of wings and tail in flight that otherwise would be too quick to be seen. Three flash heads were used, hence the three points of light reflected in the bird's eye. The Crimson Rosella inhabits coastal Queensland, New South Wales, Victoria and eastern South Australia.

Bottom:

Red-capped Parrot, *Purpureicephalus spurius*
Found only in the forests of southwestern Australia, the Red-cap is one of Australia's most richly coloured parrots, with purple, crimson, yellow, green and blue in the plumage. It almost invariably selects a hollow in a high branch, necessitating a photographic hide forty or fifty feet above the ground; consequently it has rarely been photographed in the wild. This male Red-cap, flying from his nest hollow, has just fed the five young with seed collected from the paddocks of dry grass that can be seen through the trees in the distant background.

bower at which he sings and displays during the breeding season. Most are found in jungle country, but three Australian species, the Great Bowerbird, the Spotted Bowerbird, and the Western Bowerbird, inhabit open savannah-woodlands of the north, and the near-desert country of the interior.

Despite their tremendously different environment, the desert-dwelling bowerbirds share many of the patterns of behaviour of their close relatives of the jungles. Bowers of sticks are built and 'painted', and the ground decorated with an assortment of shells, bones, flowers and other items.

The most obvious difference between the bowerbirds of the moist coastal regions and those of the drier regions is in their plumage colour. Those of the humid forests have for the most part plumages of rich or bright hues, satiny blue-black, or gold, or orange-and-black. But the inland bowerbirds are grey-brown, with the only touch of bright colour being a lilac or pink neck mantle.

Among the kingfishers of Australia are some which have found a way of life far from the 'traditional' kingfisher environment of lake and stream. Those kingfishers that are still closely tied to water are the species that occur around the tropical northern, north-eastern and eastern coasts. It is not surprising to find that these species occur in similar habitats in New Guinea.

Best known of these water-loving kingfishers is the Azure, which is found always near creeks, rivers, esturine shores or mangroves. Other species closely associated with water are the Little Kingfisher, which in Australia is restricted to coastal Northern Territory and north Queensland, and the Mangrove Kingfisher, found around the northern Australian coastline and throughout the South-East Asian region.

Three others of our kingfishers, while not closely tied to water, are confined to the humid forests near the coasts. Both the White-tailed and Yellow-billed kingfishers are confined to the tropical forests of north Queensland; both occur also in New Guinea and surrounding islands. The Forest Kingfisher, which inhabits our coastal forests, is also found in New Guinea.

By way of contrast, the remaining four species of Australian kingfishers have ventured far from the streams and moist coastal forests. The Laughing Kookaburra (a giant kingfisher) ranges well inland in eastern Australia, favouring savannah-woodlands, while the Blue-winged Kookaburra inhabits some very arid country particularly in north-western Australia. The Sacred Kingfisher, while inhabiting coastal districts, is found also far into the interior where it prefers the tree-lined watercourses.

But the kingfisher that has had the greatest success in making use of the very extensive desert and semi-arid environments of Australia is the Red-backed, which inhabits mallee, mulga scrub,

Top:
Cockatiel, *Nymphicus hollandicus*
With wings a blur of rapid action against the sky and tail fanned wide to act as a brake to check its speed, a male Cockatiel comes in to land on a broken limb of the dead tree in which its nest hollow is situated. The Cockatiel occurs in all Australian states, in open country, principally in the interior. It is a nomadic species, wandering in search of the seeds of grasses and herbaceous plants which make up its food supply.

Bottom:
Major Mitchell Cockatoo,
Cacatua leadbeateri
With crest raised, on a limb beside its nest hollow.

savannah, rocky desert ranges and sand dune country. It occurs right across the continent, and avoids only the wet coastal districts.

The fairy wrens of Australia, like the kingfishers, have evolved species to occupy almost every available habitat. Their colourful plumages are one of the delights of the Australian bush; because many have blue in the plumage they are often collectively referred to as 'blue wrens'. For sheer spectacular brilliance of colour the male Splendid Wren, seen with full sunlight on his iridescent purple-blue plumage, is a wonderful sight. Equally breathtaking is the Red-backed Wren, whose scarlet-and-black plumage resembles a flame among black coals. At the other end of the scale, the emu-wrens, mouse-like in size and movement as they creep through the undergrowth, are not so brightly coloured, but have marvellous long filamentous tails that resemble emu feathers.

The Australian wrens have diversified in choice of habitat. Some, like the Lilac-crowned Wren, prefer swampy cane-grass thickets, while the Blue and White Wren and the Rufous-crowned Emu Wren inhabit the most arid of our desert country.

Australia is the poorest of all the continents in waterfowl, having only 19 of the world's 148 species of web-footed aquatic birds. However some of the Australian waterfowl are of particular interest for their success in adapting to dry regions that would seem unsuitable for them because of long and frequent droughts. The Grey Teal, for example, accomplishes this by breeding immediately after good rain at any time of the year, thereby completing its nesting before drought conditions return, and like most inland water birds, travelling great distances to find favourable seasonal conditions.

Turquoise Parrot, *Neophema pulchella*
Caught by camera and high-speed flashes an instant before it landed on top of the stump inside which its young were hidden, this male Turquoise Parrot shows brilliant yellow under-surface, deep blue wings, and the turquoise head for which it is named. This is an uncommon species, with a localized distribution in parts of inland Victoria, New South Wales and southern Queensland.

Above:

Azure Kingfisher, *Alcyone azurea*

A tiny Azure Kingfisher, with a small fish in its bill, comes in for a high-speed landing on a perch overhanging a river bank. These birds have a very fast, darting flight, skimming a few inches above the surface of the water so that, to the eye, they are no more than a flash of blue and gold, then suddenly swooping up to land on some perch over the river. At the instant when its flight was frozen by the electronic flash the kingfisher's body was in a vertical position, so that its entire undersurface and widespread wings together formed a parachute or air brake to slow its speed, just before its outstretched feet struck the perch. To protect against any sharp twig at this moment the semi-transparent nictitating membrane has flicked across the bird's eyes.

Opposite:

Perched on a branch overhanging the river the Azure Kingfisher pauses to beat its prey against the wood before flying down to its nest tunnel beneath the overhanging river bank. Azure Kingfishers are found always near water of creeks, rivers or mangrove swamps, and inhabit northern, eastern and south-eastern Australia. This pair had chosen a stream flowing through a rainforest in north-eastern New South Wales. The surrounding jungle trees are reflected in the blue-grey river, while the gloomy overhanging river bank opposite makes a black shadow beyond the bird's mossy perch.

Black-faced Flycatcher, *Monarcha melanopsis*
Diving down from the jungle canopy to its camouflaged mossy nest, a Black-faced Flycatcher presents a fascinating pattern of shining pearl-grey wing and tail plumage. Only with the development of modern high-speed photographic techniques utilizing an exposure of 6,000 of a second, together with transistorized electronic controls is it possible to see clearly and in perfect detail the way in which a bird accomplishes the miracle of flight. Here is revealed the shape of outspread wings and tail, the interlocking of thin but strong feathers to provide the broad air-catching surfaces, and the twisting of the tail to give directional control.

At its beautifully constructed nest the Black-faced Flycatcher, with wings still raised after landing, shows the delicate greys and orange-buff underparts of its plumage. This bird of the heavy scrubs and thick forests inhabits coastal eastern Australia from Cape York almost as far south as Melbourne. Its food consists of insects of all kinds, taken from foliage and branches, or captured on the wing.

Rainbow Bird, *Merops ornatus*

During summer months in southern Australia the brightly plumaged Rainbow Birds attract attention with shrill calls and acrobatic flight as they twist and turn in pursuit of flying insects, their orange wings flashing in the sun against blues and greens of sky and forest background. Rainbow Birds, which winter in the far north, spend most of their time perched on dead limbs above paddocks or forest clearings, from which vantage points they dart after any passing bee, dragonfly or other large insect. Once captured their prey is struck repeatedly against the wood before being carried down into the underground nest where there may be as many as seven young. The nest tunnel is drilled two or three feet into soft soil; most of the digging out of loosened soil is done with the feet, but occasionally a larger stone will be carried out in the bird's beak.

Above:

New Holland Honeyeater, *Meliornis novaehollandiae*

From its perch on a stiff strong banksia flower spike, a New Holland, or White-bearded, Honeyeater investigates the massed rows of flowers, probing among them for nectar and insects. Unopened flowers are pink, turning yellow as they open out to reveal the pollen. As this happens first to the bottom rows, a wave of yellow spreads upwards until the whole of the flower spike is golden. The New Holland Honeyeater is a very widespread species, ranging from south-eastern Queensland to Victoria, southern South Australia, Tasmania and south-western Australia. The wildflower, commonly called 'Firewood Banksia', is a small tree that grows abundantly on the sandy coastal plain around Perth.

Opposite:

Western Spinebill, *Acanthorhynchus superciliosus*

Flying down to land on the red-felted stem of a Red-and-green Kangaroo-paw, a male Spinebill carries an insect it disturbed at the flowers and captured after a short chase. On the Spinebill's head is a smudge of golden pollen from the finger-like anthers of the flowers. To reach the nectar at the bottom of the green perianth tube the bird must thrust its long beak into the deep slit-shaped opening. This action brings the crown of his head in contact with the hanging thread-like style, leaving a few grains of pollen adhering to its sticky tip, and then against the row of yellow anthers which apply a fresh dusting of pollen. In this way these small birds, flying from plant to plant through a clump of kangaroo-paws, are forced to pollinate the flowers in return for the nectar taken. Both the Spinebill and the Red-and-green Kangaroo-paw are native to south-western Australia.

Top:
Yellow-plumed Honeyeater, *Meliphaga ornata*
Dashing away from the stem above its nest this honeyeater shows its yellow-edged flight feathers as the wings extend in a powerful takeoff downbeat. The bird's name, 'Yellow-plumed', refers to the yellow markings on the side of the head in the vicinity of the ear. This is a common bird of the mallee districts of north-western Victoria, south-western New South Wales, South Australia, and south-western Australia where it also inhabits the wandoo forests. The nest is a delicate basket of woven grasses suspended from slender twigs of a tree or shrub.

Bottom:
White-cheeked Honeyeater, *Meliornis niger*
Although very similar to the New Holland Honeyeater, the White-cheeked has a considerably more attractive plumage, the most noticeable difference being the large pure white cheek plumes. It inhabits heathlands and open forest country of coastal eastern Australia and south-western Australia.

Above:
Kookaburra, *Dacelo gigas*
Australia's giant laughing kingfisher is one of its best-known birds—one of the 'popular' birds, but far from popular with those who know it well. Like the small kingfishers it is a predator, taking any helpless or small creature it can find, including nestlings of smaller bushland birds. The Kookaburra inhabits open forest country throughout eastern and south-eastern Australia, and has been introduced into south-western Australia and Tasmania.

Nankeen Kestrel, *Falco cenchroides*

One of Australia's smaller birds of prey, this is one of the most attractive with its rusty-brown back and buff-white underparts. This falcon occurs throughout Australia in open country, plains and lightly timbered lands, where it is often to be seen hovering, maintaining the one position in the air perfectly, scanning the ground, then dropping suddenly to take some small creature in its talons. Here the Kestrel is shown landing at its nest, a deep hollow about twenty-five feet above the ground, where the two young, fully-feathered and ready to fly, flap their wings with excitement at their parent's approach with food.

Above:

Flame Robin, *Petroica phoenicea*

Nesting in a Snow Gum near the summit of Mt Kosciusko, these Flame Robins have built into a crevice between the tree trunk and a projecting piece of bark. The Flame Robin is a migratory species, with considerable movement taking place just before the breeding season; in summer it becomes common in the Australian Alps at altitudes above 4,000 feet, where nesting is delayed until late in the season—this nest on Kosciusko was found in January. There is also a regular movement of these birds across Bass Strait to and from Tasmania. The nest, like those of other robins, is neatly constructed, the strips of bark, fern and grass being bound together with spiders' webs and the exterior shaped to blend into a fork or fit the cavity chosen. The Flame Robin is restricted to south-eastern Australia and Tasmania, and may easily be distinguished from Scarlet and Red-capped robins by its grey rather than black upper parts.

Opposite:

Spotted Pardalote, *Pardalotus punctatus*

A male Spotted Pardalote, with wings outstretched an instant before diving into the entrance of its nest tunnel in a mossy creek bank, shows its symmetrical pattern of white markings and fiery tail. The tunnel, little more than an inch in diameter, extends twelve to twenty-four inches into soft soil to a chamber where a domed nest of shredded bark is constructed. The Spotted Pardalote takes insects from the foliage of the forests of eastern, south-eastern and south-western Australia, and Tasmania.

Above:

Scarlet Robin, *Petroica multicolor*
With wings thrown wide and tail twisted in a last-instant correction of direction, a male Scarlet Robin drops down to the rim of his bark nest in a fork of a banksia tree. The young are old enough to see their parent's approach, and beg noisily for the insect he brings. The five-inch-long Scarlet Robin may be found in forests and woodlands of all Australian states including Tasmania.

Opposite:

Red-capped Robin, *Petroica goodenovii*
An instant after landing on a twig beside his nest this male Red-capped Robin still has one wing thrown up; the other wing, half folded, is almost hidden behind his body. This robin inhabits the drier areas of all states except Tasmania; it is also absent from tropical northern Australia. The nest is particularly neatly constructed. Strips of bark, dry grass, pieces of moss and lichen are matted together with cobwebs and the small cup-shaped interior is warmly lined with fur and feathers.

Above:

Red-winged Wren, *Malurus elegans*

On his way to his nest a male Red-winged Wren alights on gumnuts among the ferns of the forest floor; the chestnut-coloured shoulder patches which give him his name are largely hidden by the raised position of his wings. Red-winged Wrens favour dense vegetation along streams flowing through the forests of the higher rainfall parts of coastal south-western Australia. Two rather similar birds, the Variegated Wren and the Blue-breasted Wren, inhabit the drier inland and northern areas. All species of chestnut-shouldered wrens build a hooded nest with a side entrance; in the case of the Red-winged Wren the nest is usually extremely well hidden in the tangled vegetation and debris of the forest undergrowth.

Opposite:

Yellow-breasted Sunbird, *Cyrtostomus frenatus*

A female Yellow-breasted Sunbird hovers for a moment before landing at the entrance of her suspended nest. The male of this species is even more colourful with a deep blue throat patch. This long nest, which has a little hood over its side entrance, is constructed of bark fibres, dry grass, and dead leaves bound together with cobwebs. Beneath the softly lined egg chamber hangs a long tail, and the whole structure is hung from a twig by a slender strap-like extension of the domed roof. Pieces of forest litter bound to the exterior make it most inconspicuous. These nectar and insect-eating birds are found in or near rainforests of coastal north-eastern Queensland, as well as in New Guinea, the Celebes, Solomon and other islands. Sunbirds are somewhat like honeyeaters, but even more closely resemble hummingbirds, often hovering in front of a flower that is not easily accessible from a perch.

Above:
Banded Wren, *Malurus splendens*
This male Banded or Splendid Wren is about to
land on a flower spike of a grass-tree with an
insect captured nearby. In the background is a
twisted creamy-coloured flower spike of a
blackboy plant. The fully-plumaged male of
this species is one of the most resplendent of
Australia's small birds, with its brilliant violet-
blue being broken only by the black bands
around breast and nape of neck. The Banded
Wren inhabits south-western Australia in low
scrub and open forest country. Its nest is of the
usual wren pattern, with a domed top and a
side entrance; it is placed in a bush usually
one to four feet above the ground.

Opposite:
Blue-and-white Wren, *Malurus cyanotus*
Found across Australia from western Queensland, New South Wales and Victoria through
South and Central Australia to Western Australia, the Blue-and-white Wren inhabits the
driest country, and favours scrubby sandplains
and spinifex, and the samphire flats around salt
lakes. The male's white shoulder patches are
conspicuous at a distance. When he is flying,
his dark blue body is often inconspicuous against
blue-green vegetation so that the flickering
white wing patches at first glance could be
mistaken for a large white butterfly in flight.
The domed nest is usually close to the ground in
a small bush or clump of grass. Like most other
small birds of the interior, this wren will nest
after good rains at any time of the year.

Above:

Emus, *Dromaius novaehollandiae*
A pair of Emus come to drink at a bush pool, one
remaining standing and wary while its mate
drops to a position that enables it to scoop up
the water more easily. The Emu is the tallest
of Australian birds, reaching a height of six feet,
but it is not quite as heavy as the more solidly
built Cassowary. Once occurring almost
throughout Australia, Emus are now absent
from most settled areas but remain common on
the vast dry inland plains.

Opposite:

Blue-breasted Wren, *Malurus pulcherrimus*
This is one of the chestnut-shouldered group of
'fairy wrens', looking very similar to the Red-
winged Wren. The male Blue-breasted Wren
has a slightly lighter blue on his chest and throat
than the male Red-winged; there are other
slight differences of plumage, and the songs of
the two species are quite distinctive. This wren
occurs on the Eyre Peninsula of South Australia,
and in south-western Australia.

Opposite:

Superb Lyrebird, *Menura superba*

At the climax of the Lyrebird's display the long tail curves forward over his head so that he is completely enveloped in its shimmering, silvery filaments. At the same time he sings, imitating in rapid succession the calls of many of the birds of the forest. The male Lyrebird devotes all his energies to his displays at the mound during the winter breeding season, while the females alone build the nest and attend the young. Superb Lyrebirds occur in forested south-eastern Australia from coastal Queensland to Victoria, and have been introduced into Tasmania.

Above:

Black Swan, *Cygnus atratus*

One of the paradoxes of a land of unusual creatures, the Australian swan is black instead of white, accentuating the brilliant red of the bill. In flight, or whenever the wings are outstretched, the white primary and secondary flight feathers are in striking contrast to the black. Black swans may be seen in almost any part of Australia and Tasmania, but are most common in the south-east and south-west.

Above:

Orange Chat, *Epthianura aurifrons*

The Orange Chat is a desert-loving species which may be encountered in some of the most arid parts of Australia. It prefers the samphire margins of salt lakes where its nest is extremely well hidden in a saltbush or some other dense low shrub within a few inches of the ground. The finding of these nests is made difficult by the behaviour of the birds. The chats do not fly direct to the nest but land on some other shrub before running along the ground, keeping out of sight, to reach the nest. These confusing tactics are repeated in reverse when the birds leave the nest. In this open type of terrain, where there seems always to be a crow or hawk lurking somewhere in the distance, this evasive behaviour must have considerable survival value. Here the male Orange Chat brings a grasshopper to a nest where the young are almost old enough to fly.

Opposite:

Crimson Chat, *Epthianura tricolor*

Occurring throughout inland Australia from western Queensland to Victoria and through Central Australia to inland Western Australia, the Crimson Chat favours open country with low bushes, where it spends most of its time on the ground looking for insects. The species is highly nomadic, being seen in great numbers in any part of the interior following good rains, and being completely absent from districts that are excessively dry. The nest is well hidden in a low bush, often within several inches of the ground; these dense violet-flowered mulla-mulla bushes are favourite sites in some regions.

72

3 The Furred Animals

Native-cat, *Dasyurus geoffroii*
When caught late at night by hidden camera and flash, this native-cat was about to tear apart its prey, a Red-capped Parrot. The native-cats, of which there are four species, are skilful climbers, spending most of their time in the treetops. The largest, known as the Tiger Cat, is restricted to the forests of the east coast from Queensland to South Australia and Tasmania; the three smaller species inhabit eastern, northern and south-western Australia, so that most parts of the continent have at least one species of this predatory marsupial in their fauna. The native-cats have up to eight young in a litter, and these cling to the fur of their mother for several weeks after leaving the pouch.

Of all Australia's living creatures, no segment is more distinctively Australian than its marsupials. While the monotremes (Platypus and Spiny Anteater) seem to represent an older, more primitive type of mammal than found anywhere else in the world, it is the marsupials, with their bewildering array of species, that dominate Australia's terrestrial fauna.

In size the marsupials range from insectivorous planigales much smaller than any mouse, to the great Red Kangaroo which stands five and half feet tall and weighs up to 180 pounds.

Marsupials differe from all other furred animals in their way of reproduction. Outwardly, the possession of a pouch sets them apart, but in many species this is not readily discernible. However all marsupials are fundamentally different in that they have never evolved the placenta, the structure within the womb which nourishes the embryonic young. Higher mammals, which produce large, almost fully-developed young, have the placenta, and are known as 'placental mammals'.

But the young of marsupials must be born at an incredibly early stage, and must find their way unaided to the pouch region, where they complete their growth firmly attached. The young of a large kangaroo is, at birth, only three-quarters of an inch in length, and only its fore-paws, with which it struggles instinctively through the fur, have developed to anywhere near their final shape.

The Australian mammal fauna as a whole is typical of an island rather than a continent in that its 225 species belong to only four orders—monotremes, marsupials, rodents and bats.

The monotremes are considered to be relics of ancient origin, and may have always been restricted to the Australian region. The two species, Platypus and Spiny Anteater, are primitive in their similarity to the reptiles from which they have evolved, especially in their reptile-like egg-laying way of reproduction.

At the other extreme, the most advanced mammals, the placentals, give birth to comparatively larger young which in some cases (fawns, foals) can almost immediately run with their parents. So the stage reached by the marsupials is something of a mid-point between the egg-laying monotremes and the placental mammals.

Australia's marsupials can be considered in four major groupings—the marsupial carnivores, the kangaroos and wallabies, the possums wombats and Koala, and the bandicoots. Each of these groups has evolved to occupy a place in the Australian environment which on other continents is occupied by an entirely different animal.

Many of these Australian animals, although of another origin, have over the millions of years come to resemble animals which follow a similar way of life in other parts of the world. This tendency for animals of different origins to acquire similar characteristics is known as 'convergent evolution'.

We have in Australia the Tasmanian Tiger or Thylacine with much the same appearance as the unrelated wolf of Asia and north America; each has evolved independently to the wolf-like shape because it is apparently the best mammal pattern for this particular way of life.

In Australia we have tree-climbing, carnivorous marsupials, which we call native-cats because we see in them a considerable resemblance to the cats of the other continents. But these marsupials are not related to cats. They and the true cats have evolved a similar shape because this is the best form for an arboreal, nocturnal, predatory mammal to take.

But not always do animals of similar habits acquire such similarity of appearance and movement. Australia's large herbivorous mammals, the kangaroos and wallabies, are very different to the grazing animals, the deer, antelope, horses and so on, of other parts of the world. For some reason the large Australian herbivores have chosen the hopping rather than the running way of fast movement. Possibly hopping is advantageous for an animal that must constantly avoid obstacles on the ground, such as the clumps of prickly porcupine grass, boulders and logs which are common in the Australian plains or forests.

But even in this case, convergent evolution has produced similar head and jaw structures, suitable for grazing.

In Australia, possums and cuscuses take the place of monkeys, and gliding phalanges such as the Sugar Glider resemble the gliding squirrels of other lands. The Platypus is rather like the beaver, both being mammals adapted for an aquatic life, and sharing such characteristics as the flattened, paddle-like tail, even though they are of quite different origins.

Each major marsupial group contains many animals of special interest. Although some species occur also in New Guinea, most are endemic to Australia.

The marsupial carnivores were probably the original marsupial inhabitants or immigrants to Australia; it is thought that all other kinds evolved from them.

These carnivorous marsupials vary in size from the wolf-like *Thylacine* down through many superficially cat-like and rat-like species to the tiny insectivorous marsupials, some of which are much smaller than a mouse.

The flesh-eating marsupials of cat size are commonly called 'native-cats'. Although they are cat-like in their hunting prowess and ferocious attack, and somewhat cat-like in appearance, they are not related to the ordinary feline cats, which have been introduced into Australia.

The carnivorous marsupials have shallow depressions or pouch areas, instead of the deep pocket-like pouches of kangaroos and

Red Kangaroo, *Megaleia rufa*
The largest living marsupial, the Red or Plains Kangaroo has an extremely wide distribution across the continent wherever there are plains, preferring dry regions of less than fifteen inches rainfall; it does not extend into tropical northern grasslands. The males may grow to a huge size, weighing as much as 180 pounds, but females are much smaller and do not often exceed 60 pounds. Fur colour is variable; the females which are generally a different colour to males are sometimes called Blue Fliers or Blue Does. The joey at birth is incredibly small compared with the bulk of the adult kangaroo, being only three-quarters of an inch in length; yet it can crawl unaided through the fur to the pouch. Not until some 190 days have elapsed does the young kangaroo make another appearance. From that date it begins to leave the pouch for short periods, gradually spending more and more time out, until aged about 235 days it climbs from the pouch for the last time.

others, and here the young cling tightly in the fur. After leaving this rudimentary pouch they hang on all over their mother, who staggers along under the weight of six or eight half-grown young native-cats.

Native-cats, of which there are four species including the large Tiger Cat, are brownish or ginger with white spots. They are skilful climbers but hunt on the ground as well as in the trees, preying upon birds, insects and small mammals.

The many rat-sized and mouse-sized carnivorous marsupials can be positively distinguished from true rats and mice by their needle-pointed, cat-like teeth, compared with the chisel-shaped incisors of the rodents. In general appearance the small carnivorous marsupials have an alert, sharp-faced appearance quite different from the blunt-nosed rodents.

The smallest of the marsupial carnivores are the insectivorous planigales, or flat-headed marsupial mice. These are the smallest of all marsupials, three of the four species being much smaller than a mouse. For their size they are ferocious little animals, killing grasshoppers and other creatures almost as large as themselves.

Only slightly larger than the planigales are the dunnarts or narrow-footed marsupial mice. These mouse-sized predators are principally insectivorous, but occasionally have been known to kill small lizards, small birds, and ordinary mice.

The Fat-tailed Dunnart, like a miniature native-cat, will when frightened give a savage-looking display of wide-opened mouth and rows of needle-pointed teeth. The tail of this species becomes thickened when food is abundant. As its range extends into the semi-desert country, the food stored in the tail is most valuable in times of drought. In addition it is able to lower its body temperature when sleeping, going into a state of torpor, and thereby conserving food.

But of all the carnivorous marsupials one of the most highly specialized is the Numbat or Marsupial Anteater, which lives almost entirely upon termites. For this way of life it has acquired a tremendously long thin tongue, and its teeth have become very much reduced.

Kangaroos, the family Macropodidae, are divided into about thirty species, and occur only in Australia and New Guinea. They vary in size from the very large Red, the Grey Kangaroo and the Euro through many intermediate-sized wallabies and pademelons to the little rat-kangaroos, some of which are scarcely as big as a rabbit. Among them are kangaroos adapted for life in the treetops, and others that inhabit places of cliffs and boulders on the tops of rugged ranges.

The possums, the Koala and the wombats can all be separated from any of the kangaroos, quite apart from general

Agile Wallaby, *Macropus agilis*
The Agile Wallaby, known also as Sandy Wallaby and Jungle Kangaroo, is one of the commonest small kangaroos of tropical northern Australia, where it feeds on the extensive grassy plains and shelters under the heavier vegetation of the jungle-like monsoon forests or other dense vegetation. The distinct white cheek and hip stripes, and sandy-brown or golden colour help recognition of this species.

appearance, by their hind feet, which without exception are hand-like, and quite different from the slender hind feet of all kangaroos, wallabies and rat-kangaroos.

The common Brush-tailed Possum, the two species of cuscus, the Scaly-tailed Possum and the various ring-tailed possums are comparatively large; at the other end of the scale are the mouse-sized pigmy possums and the highly specialized, nectar-eating Honey Possum.

There are five species of gliding possum, from the cat-sized Greater Glider, the Yellow-bellied Glider, the Squirrel Glider, the small Sugar Glider, and the tiny Feather-tailed Glider.

The Koala, once regarded as some sort of big plump tail-less possum, is now known to be more closely related to the four species of wombats than to any other group. It is one of the most specialized of Australian animals, being almost exclusively arboreal and living on the leaves of just a few species of smooth-barked eucalypts.

Bandicoots, rather small, terrestrial nocturnal marsupials, dig in the earth and leaf-litter in search of insects and other small invertebrates. Most of the bandicoots are sharp-nosed, rat-like animals, but two are of special interest. The Rabbit-eared Bandicoot, known also as the Dalgyte or Bilby, is one of the most beautiful of Australia's native animals, with soft silky fur, long black-and-white tail, slender pointed nose, and very large long ears.

Once common in parts of Australia, both east and west, the Dalgyte is extinct except in some remote and desert areas.

The other strange member of this group is the Pig-footed Bandicoot, a graceful little animal slightly smaller than a rabbit, with long ears, long tail, and slender, deer-like legs that terminate in toes resembling the cloven hoofs of a pig. This species is probably now extinct, the last having been seen in Central Australia almost fifty years ago.

Top:
Western Grey Kangaroo, *Macropus fuliginosus*
Also known as the Black-faced Kangaroo, Mallee Kangaroo or Sooty Kangaroo in various regions, the Western Grey inhabits forests and woodlands of south-western New South Wales, parts of Victoria and South Australia, and south-western Australia. It may be distinguished from the Great Grey or Forester by its colour, light grey-brown to dark brown rather than silvery grey.

Bottom:
Yellow-footed Rock-wallaby, *Petrogale xanthopus*
Rock-wallabies shelter during the day in caves and crevices of rocky and mountainous country where they are safe from most potential enemies. Among the boulders and on the rocky slopes they show amazing agility and precision as they leap effortlessly in places where a slip or error in judgement would mean a long fall. The rock-wallaby group is widely distributed, and has seven species; the Yellow-footed, also known as the Ring-tailed Rock-wallaby, occurs in suitably rugged habitats from south-western Queensland through western New South Wales to the Flinders and Gawler Ranges of South Australia.

Koala, *Phascolarctos cinereus*

One of the best known and most highly specialized of Australian animals, the Koala inhabits suitable areas of forest and woodland from south-eastern Queensland through eastern New South Wales and Victoria to south-eastern South Australia. Fossil remains show that it once occurred in south-western Australia. The Koala is almost exclusively arboreal, coming to the ground only occasionally to cross to other trees; it will eat only the leaves of a few species of smooth-barked eucalypts. For this treetop life the Koala has developed long arms, sharp claws, and a powerful grip between opposing fingers, giving it a strong hold on even a smooth tree trunk. After a young Koala leaves its mother's pouch it rides upon her back for several months before finally becoming independent.

Above:

Pigmy Possum, *Cercartetus concinnus*

The South-western Pigmy Possum is a mouse-sized marsupial with a clinging prehensile tail and tiny feet that are like hands both in their appearance and their ability to grasp for climbing. Australia has five pigmy possum species; this one occurs in the eucalypt forests of western Victoria, southern parts of South Australia, the extreme south-west of New South Wales, and in Western Australia. The pigmy possums feed on nectar of flowering trees, and insects which they pounce upon with surprising speed, then hold in the forepaws to eat daintily, discarding legs, wings and hard portions. Like most small marsupials, pigmy possums are nocturnal, hiding by day in old birds' nests, in hollows, in dead grass-trees or in crevices under bark. Like the larger marsupials the pigmy possum has a pouch, in which as many as six tiny young are carried.

Opposite:

Short-eared Brushtail, or Mountain Possum, *Trichosurus caninus*

Inhabiting the mountain rainforests and heavier eucalypt forests of south-eastern Queensland southwards to eastern Victoria, this is a distinctive species which can be distinguished from the Common Brushtail by its short rounded ears. It is a large, powerfully built possum, with long dense fur ranging in colour from grey to dark brown. Like other possums the Short-eared Brushtail is nocturnal, sheltering by day in hollow trees.

Opposite:

The Dingo, *Canis familiaris*

In its most common form the Dingo is tawny yellow with paler undersurface and, frequently, white feet and tail tip. But there is considerable natural variation of coat colour, quite apart from any changes that may have been brought about by crossing with domestic dogs. The Dingo is a quite distinctive breed of dog, with his own peculiar forms of behaviour, and a howl that is all his own. It is thought that the Dingo was brought to Australia from Asia by the Aborigines, perhaps 6,000 years ago. The Dingo was probably one of the reasons for the extermination of the Thylacine and the Tasmanian Devil from the mainland of Australia. In spite of the killing of many hundreds of thousands of dingoes in pastoral areas, the Australian native dog is still common in many wild or remote regions.

Above:

Spiny Anteater or Echidna, *Tachyglossus aculeatus*

One of the world's two living species of monotreme, the Spiny Anteater shares with the Platypus the distinction of being a primitive link between the reptiles and the mammals. The Anteater and the Platypus are the only mammals to lay eggs, and have reptile-like features of reproductive system, skeleton and body temperature regulation. Although the Spiny Anteater is very widespread, occurring in a variety of habitats from rainforest to semi-desert, it is not often seen. When disturbed its immediate reaction is to dig vertically downwards, and once half-buried, is extremely difficult to dislodge.

Above:

The Platypus, *Ornithorhynchus anatinus*

Probably the strangest of all Australian animals, this creature at first appears to be a peculiar mixture of bird and mammal. But such unusual features as the duck bill and webbed feet are adaptations to the amphibious way of life, while egg-laying rather than live birth is a character- istic retained from its not-so-distant reptile ancestors. The Platypus usually emerges from its burrow to feed in late afternoon and early morning. Its sensitive bill locates by touch the worms, tadpoles, crustaceae, insect larvae and other aquatic creatures upon which it feeds; eyes and ears are closed when it is submerged. The Platypus has remained quite common in freshwater lakes and streams of eastern Australia and Tasmania.

Opposite:

Western Ringtail Possum, *Pseudocheirus peregrinus*

This dark-furred ringtail possum is an isolated western race of the Common Ringtail. Cut off from all other ringtail possum populations by deserts, the ringtails of the south-west have gradually acquired a very dark colour. Ringtail possums have a long prehensile tail which is often carried rolled in a spiralling ring. The tail can be used almost like another hand in grasping branches or foliage. The attractive, rather shy and gentle ringtails have always a fixed, staring expression; their big rounded eyes are so adapted to darkness that they seem to find daylight distinctly uncomfortable.

Above:

Spotted Cuscus, *Phalanger maculatus*

An inhabitant of the rainforests of Cape York, the Cuscus, with its very short ears, thick woolly fur and grasping tail, is quite monkey-like in appearance. The Spotted Cuscus and the rather similar Grey Cuscus are marsupials of New Guinea origin. Both species occur in New Guinea and other northern islands.

Opposite:

Fat-tailed Dunnart, *Sminthopsis crassicaudata*

Australia has eleven different small marsupial-mice known as dunnarts; of these, five have peculiar thickened tails which make their group unmistakeable. This particular species of *Sminthopsis* occurs in woodlands and grasslands from south-western Queensland through western New South Wales and Victoria to South Australia, and also in south-western Australia. The Fat-tailed Dunnart inhabits rather dry regions, where the fat storage of the thickened tail could be of survival value at times of drought or food scarcity for any other reason. At times the tails of specimens caught in the wild may be quite thin. The dunnarts nest in hollow logs on the ground or under stones, where they may sometimes be found in a torpid condition, almost incapable of movement for some time.

Above:

Green Ringtails, *Pseudocheirus archeri*
Their fur greenish-brown like the mossy rainforest trees of their north-eastern Queensland haunts, a mother and her half-grown young cling to a jungle branch. These beautiful possums, which have prominent light-brown eyes, white ears and undersurfaces and two stripes down the back, are found only in the high altitude rainforests of the Atherton Tableland and adjacent mountains.

Opposite:

Honey Possum, *Tarsipes spenserae*
The Honey Possum, barely the size of a small mouse, is a marsupial which over millions of years of evolution has acquired special adaptations for feeding on the nectar of flowering trees and shrubs. It has gained a long pointed snout which it uses like a honey-eating bird and it has almost lost all teeth, these being of little use for a nectar-eating animal. All four feet are like tiny hands, most suitable for climbing. Identification of the Honey Possum is made easier by the three distinct stripes down its back. The Honey Possum is quite unique, there being no other animal quite like it anywhere else in the Australian fauna; it is found only in the south-western corner of the continent where the abundance of wildflowers has made possible the evolution of so specialized a creature.

Above:

Feathertail Glider, *Acrobates pygmaeus*

Clinging to the flowers of a north Queensland tree, the tiny Feather-tailed Glider looks in the direction of camera and flash, his bulging black eyes exploring the darkness and his sensitive ears picking up each faint sound. The flash of light which recorded his image on film is reflected as white circles of light in each eye. The Feather-tailed or Pigmy Glider has flap-like gliding membranes which stretch from out-spread arms to legs, enabling it to make descending gliding leaps from branch to branch or tree to tree. This marsupial inhabits forests and woodlands from north-eastern Queensland to south-eastern Australia.

Opposite:

Sugar Glider, *Petaurus breviceps*

A tough little insect and nectar-eater, the Sugar Glider inhabits eucalypt forests and woodlands under climatic extremes ranging from the tropical heat of the Kimberleys, the Northern Territory, and northern and eastern Queensland, to the cold of the snow-covered slopes of the Australian Alps. The Sugar Glider is considerably larger than the Feather-tailed, being about fifteen inches from nose to tip of long tail. The flying membranes are loose sideways extensions of the body skin out to wrist and ankles, making a broad enough surface to permit glides up to fifty yards when the takeoff point is the top of a tall tree. Sugar Gliders usually nest in hollows of trees, where a nest of leaves is constructed. The young are carried in the pouch for several months, then left in the nest while the mother is out after food.

Above:

Hairy-nosed Wombat, *Lasiorhinus latifrons*

Although superficially resembling the Common Wombat, this species differs in having its muzzle covered with fine silky hairs instead of being bare, while its fur is beautifully soft and silky. This is a slightly smaller wombat, growing to about three feet in length and weighing 60 to 70 pounds. Wombats are extremely strong, and powerful diggers. The burrows of the Hairy-nosed Wombats are grouped together to form a warren often with a huge pit as the common entrance. Wombats are marsupials, and have a pouch which opens backwards. Hairy-nosed Wombats are now restricted to South Australia, with colonies on the Nullarbor Plain, along the western banks of the Murray, and in smaller colonies on the Yorke and Eyre Peninsulas.

Opposite:

Numbat, *Myrmecobius fasciatus*

One of Australia's most beautifully-coloured marsupials, and one which has become extinct in eastern Australia, the Numbat still survives where patches of suitable woodland habitat remain in the south-western corner of the continent. This almost defenceless creature has survived the introduction of the fox and the feral cat because it is not nocturnal like those predators. The Numbat is out and about in search of termites during daylight hours. At the slightest sound of danger it dashes through the undergrowth to one or other of the fallen hollow logs and limbs scattered through its territory. Once wedged deep inside the narrow cavity of a long and usually massive piece of rock-hard wandoo trunk or limb the Numbat is safe from most potential enemies. Man undoubtedly has been the greatest enemy of the Numbat, not by killing any of these beautiful animals, but in destruction of its woodlands habitat for farm-land. Often, where the trees have been allowed to stand in forests and various small reserves, repeated burning-off has removed all the hollow logs and sticks which shelter the Numbat and its termite food. The Numbat lives almost entirely on 'white ants', scratched from the ground among sticks and logs of the forest floor, and licked up with the enormously long tongue.

Bearded Dragon, *Amphibolurus barbatus*
When disturbed this lizard faces its foe, raises its body and opens its mouth wide, at the same time expanding its throat to form the 'beard' which makes it look a much larger and more formidable opponent. The performance must be frightening enough to deter many potential predators, although the Bearded Dragon is actually completely harmless.

Australian reptiles are a segment of the reptile fauna of the old world tropics; they are 'modern', being at a comparable stage of evolution to the reptiles of other continents. There are no relic forms of reptile still at an early stage of evolution, like the monotremes of the mammal fauna.

New species of reptiles are still being discovered; at present in Australia approximately 400 are known. There are five families of snakes, one family of freshwater tortoises, two families of marine turtles, and two species of crocodiles.

The dragon lizard family, Agamidae, has many species, including some that are colourful and of unusual shape. These dragon-like lizards have overlapping scales on the body and limbs. Their tails are long, up to three times the length of the head and body, and terminate in a fine point.

One of the strangest of this family is the Mountain Devil. This small reptile is covered with thorn-like spines which give it a most alarming appearance, though it is in fact completely harmless. It is one of the most specialized of Australian reptiles, living entirely upon very small black ants.

The Mountain Devil inhabits the arid inland, where any additional water-gathering capability is a distinct advantage. It is able to drink through its skin, thereby making full use of any brief abundance of water, such as may occur in a short thunderstorm, or heavy dew on a cold night.

The water quickly soaks all over the skin, which has a sponge-like absorbent quality, and is drawn up by capillary action through minute channels to the mouth, where it is swallowed.

The remarkable Mountain Devil (or Thorny Devil) has also the ability to change its colour, making parts of its pattern become a darker or lighter tone to match the surroundings.

Equally bizarre in appearance, but inhabiting the jungles rather than the deserts, is the Forest Dragon, which when seen close-up has the appearance of a dinosaur, with a row of saw-tooth scales down its back and a strange greenish colour.

Australia has a considerable number of geckoes (family Geckonidae) which are easily distinguished from other reptiles. They have usually a velvety appearance and are soft to touch, due to their extremely small scales (called granules) which do not overlap.

All Australian geckoes have well-developed functional limbs. Their tails, which are readily shed at times of danger, have a wide variety of shapes. Many geckoes are beautifully patterned, though not brightly coloured, because they are nocturnal, as the slit-like pupils of their large eyes indicate.

The snake-lizards and worm-lizards (Pygopodidae) make up the only family endemic to Australia and New Guinea. There are about 12 species, which have an elongated, snake-like appearance, with fore-limbs entirely absent and hind limbs extremely small or almost invisible. Unlike snakes, which have a short tail, these snake-like lizards have a tail that is much longer than the body. This tail is fragile, easily lost and replaced.

The skinks (family Scincidae) are among the most familiar of our reptiles because they are active during daylight. This large family exceeds in number of species and abundance of individuals, all other kinds of Australian lizards.

Skinks have usually a smooth and often very glossy surface. Their tails can be lost, and will re-grow. Limbs may be well developed and functional on some species, but no more than mere stump-like vestiges on others, and even entirely absent externally on some very snake-like members of the family. The heads of skinks are covered with armour-like plates rather than scales.

The goanna family (Varanidae) is represented in Australia by a single genus, *Varanus*, whose members are distributed throughout Australia. They range in size from very small goannas of seven inches maximum length, up to giants of seven feet or more.

Goannas may be distinguished from other lizards by their elongated form together with the covering of both head and body with small, close-fitting scales.

Their limbs are powerful, and their feet equipped with strong sharp claws. Although most are terrestrial, some spend much of their time on trees, while a few, the water goannas, have adopted an aquatic way of life.

Australia has about 130 species of snakes, of which 20 are sea snakes that are not restricted to Australian waters.

The land snake fauna of Australia differs from that of the rest of the world in that the majority of the venomous species belong to the front-fanged Elapidae family, including the dangerous species such as the Taipan, Death Adder and Tiger Snake.

In other parts of the world the majority of venomous snakes are in the back-fanged Colubridae family, which appears to have arrived in Australia only comparatively recently. The common Green Tree-snake is an example of this less dangerous family of snakes.

In Australia there are ten species of python (family Boidae) ranging from the extremely large Amethystine Python of tropical north Queensland to the little Pigmy Python. The Carpet Snake is a very common and very widespread member of the family.

Frilled Lizard, *Chlamydosaurus kingii*
The brightly-coloured frill which normally folds down around the lizard's neck expands suddenly, while the mouth is opened wide. The display of warning colours and pretended ferocity causes most enemies to hesitate, giving the Frilled Lizard a moment to scuttle off into thick vegetation or climb into the dense crown of a pandanus palm.

Above:

Mountain Devil, *Moloch horridus*

This bizarre little creature, of startling appearance though completely harmless, is not only extremely variable in its colouring, but has the ability to change colours to some extent to blend in with its surroundings. Usually this involves making certain parts of the colour pattern darker or lighter, of a richer or paler hue; or changes can take place within a limited range, as from dusky yellow to dull orange-brown. It is thought that changes in colour may also occur when the Mountain Devil is alarmed or excited. When disturbed, the Mountain Devil makes no attempt to escape or defend itself, but lies motionless, head flat on the ground, where it is very hard to see. If tipped upside down, it immediately rotates its tail with a very strong lever action, bringing itself right-way-up almost instantly, and thereby presenting again the thorny protective upper surface of its body. The Mountain Devil's food consists of tiny black ants, of which a thousand or more must be eaten each day. Six or seven eggs are laid in a tunnel about ten inches deep; sand is filled in over the eggs, which hatch in eighteen days or less. The *Moloch*, also called 'Spiny Devil', occurs in desert and semi-arid parts of all states except Tasmania and Victoria.

Opposite:

Forest Dragon, *Goniocephalus boydii*

Looking more like some colourful, tree-climbing dinosaur than a lizard, the Forest Dragon is appropriately a denizen of the gloomy jungles of north-eastern Queensland. In this mossy habitat the reptile's greenish colour and motionless posture on limb or tree trunk of similar colour and texture make it difficult to see. The insectivorous Forest Dragon is quite a skilful climber, though not generally fast-moving on the ground. Despite its awesome appearance it is a docile and quite harmless creature that depends upon protective colours and spiny armour-plating for its first line of defence. This is said to be an uncommon species, perhaps because it is easily overlooked, and because of the remoteness of its tropical forest habitat.

Bungarra, or Sand or Gould's Goanna, *Varanus
gouldii*
The massive Bungarra may attain a length of
five feet, and at close range resembles some huge
prehistoric reptile, particularly when it rears
up on its hind legs. Generally it is rather slow-
moving, confident that its size and formidable
claws are sufficient protection. But if need be
it can dash away at a very fast pace. The colour
pattern of the Bungarra is quite distinctive,
yellow with many black spots, darker above,
and always with a dark line through the eye.
Like other large goannas this widespread species
feeds upon carrion and any small mammals,
including rabbits, that it can catch.

Above:

Scale-footed Lizard, *Pygopus lepidopodus*
Growing to a length of about two feet, this legless lizard has a very snake-like appearance. It can be recognized as a lizard because the tail is longer than head plus body (snakes having a tail shorter by far than the body) and by the presence of ear openings and the small paddle-like remnants of hind limbs. The colour of this species is rather variable, being usually reddish, but ranging from light grey to dark brown, sometimes without dark spots. It is found in most parts of Australia.

Opposite:

Black-striped Snake, *Vermicella calonota*
Some of Australia's very small snakes are beautifully coloured and patterned; one of the best of them is the little Black-striped Snake, a burrowing species which lives largely on termites. When approached it may strike threateningly, but this is mostly bluff—although venomous, it is not considered dangerous. The Black-striped Snake is closely related to the Bandy-bandy, *Vermicella annulata*, a well-known small burrowing snake which has black and white rings around its body.

Carpet Snake, *Morelia variegata*

Gliding across a mossy boulder a Carpet Snake pauses momentarily, its tongue flickering to collect the scent of the trail it is following. The slit-like pupil of the intricately patterned iris of the python's eye, like that of a cat, can be opened very wide for best night vision; the small warm-blooded creatures that make up its prey are most active after dark. The forked tongue assists in detecting the scent of the prey.

The Carpet Snake, found in most parts of Australia (but not Tasmania) is beautifully patterned in a manner resembling some old-fashioned carpets. It grows usually to a length of eight or nine feet, though there are records of occasional specimens twelve feet long. Like all other pythons they are not venomous, the prey being killed by constriction then swallowed whole.

Above:

Thick-tailed Gecko, *Phyllurus milii*

Geckoes are small nocturnal insect-eating lizards which are quite harmless, although some species may attempt to frighten an enemy by making threatening noises or swelling their bodies. Their tails detach easily when they are handled. Geckoes are soft, unlike the usually hard, rough-textured lizard skin. The eyes are large, in keeping with nocturnal hunting ways, and the feet are sometimes equipped with dilated adhesive discs that enable them to climb across smooth surfaces. Geckoes are commonly found under rocks and pieces of bark.